Table of Content

Table of Content 1
Preface 3
Part 1 Physical Design 4
    Chapter 1 Static Timing Analysis (STA) 5
        Q1: What is metastability? 6
        Q2: What does STA do? What does being "synchronous" mean? 7
        Q3: What are setup time and hold time? 9
        Q4: What are setup time and hold time constraints? 10
        Q5: What is the benefit of using half-cycle-path? 14
        Q6: What are the sources for clock uncertainty? 15
        Q7: How does STA check reset removal & recovery / clock gating cell / data to data timing? 17
        Q8: How does STA verify async FIFO functionality? 20
        Q9: How does STA check latch based design? 22
        Q10. How does multi-cycle-path (MCP) work in STA? 26
    Chapter 2 Design Constraints (SDC) 27
        Q11: What are design / library objects? How to access & manipulate these objects? 28
        Q12: How to set single-clock design constraints in Post-CTS run? 31
        Q13: How to set I/O constraints for single-clock design in Post-CTS run? 33
        Q14: How to set multi-synchronous-clock design constraints? 37
        Q15: How to set generated clock design constraints in Post-CTS run?. 40
        Q16: How to set mutually exclusive synchronous clock design constraints? 41

Q17: How to set asynchronous clock design constraints? 44

Q18: How to verify SDC? 45

Chapter 3 STA Tool / PrimeTime 47

Q19: What is PrimeTime flow? Can you write a simple PrimeTime STA script? 48

Q20: What to check before running PrimeTime? 53

Q21: What are graph based analysis (GBA) and path based analysis (PBA)? 54

Q22: What are OCV / AOCV / POCV? 56

Q23: How to calculate timing slack using OCV? 58

Q24: What is CRPR in PrimeTime timing reports? 61

Chapter 4 Timing ECOs 63

Q25: What is the procedure for timing ECOs? 64

Q26: How to fix setup time and hold time violations? 67

Q27: What are timing ECO tools? 69

Part 2 Silicon Debug 71

Q1: What are common Design for Debug (DFD) techniques? 72

Q2: How to implement checksum? 79

Q3: How to implement Cyclic Redundancy Check (CRC)? 80

Q4: How to identify which bit of the 32b register has stuck-at fault? ... 82

Part 3 Behavioral Questions & Useful Interview Tips 85

Q1: What verbs to use in your resume to stand out? 86

Q2: Interview etiquette and best interview practices 90

Q3: How to answer "Any Questions for Me" at the end of an interview? 92

Q4: What to write in a "Thank You Letter"? 95

Q5: Why and how should you follow up with your job application? 96

Preface

As hardware engineers, we understand how stressful and struggling the hardware interview process can be. We also suffered the same pain as you might have when desiring to grow further or exploring new opportunities in the hardware industry.

Unlike software engineers who are able to find countless online resources such as LeetCode, Stack Overflow, etc., hardware engineers can hardly find their counterparts. The golden information is here and there, but nowhere summarizes it in an organized way that can be easily followed.

This series of books is intended to close the gap, by sharing our knowledge, experience and perspective towards digital design interviews. Our goal is for readers to gain real industrial experience by understanding what skills the companies are looking for. The structure of this book is organized in the same way as how the modern ASIC / VLSI industry partitions the workflow.

Specifically, in this book, we focus on Physical Design and Silicon Debug, from RTL designers' perspective. Both are important skill sets that all RTL design engineers should have some basic understanding. In addition, we share a few behavioral questions and useful interview tips at the end of this book.

We do hope you find the information in this book useful for preparing digital design interviews, and landing a dream job in the industry.

Part 1 Physical Design

Chapter 1 Static Timing Analysis (STA)

Q1: What is metastability?

We are already familiar with the term "metastability" and its association with CDC checks. But people often do not realize that preventing metastability from happening is the reason we need static timing analysis (STA) in the first place.

Metastability happens when a sequential logic has timing violations, and it enters a quasi-static state where its output settles to neither low nor high. For example, it can happen when a register (FF) has a setup or hold time violation. When setup time or hold time violation occurs, the output of that register becomes metastable.

For synchronous paths, designers should check all timing paths for setup time and hold time constraints. Designers rely on static timing analysis (STA) to perform a thorough check against all timing paths.

Q2: What does STA do? What does being "synchronous" mean?

What is STA?

Static Timing Analysis, a.k.a., STA, is timing verification methodology.

STA is exhaustive, since it uses formal, mathematical techniques instead of dynamic logic simulation, to perform analysis. STA will first identify all timing paths inside the design and all possible clock relationships, and then perform timing checks on all paths.

STA is also constraint driven, i.e., by default, it does not report a path that is not constrained for timing.

Unlike CDC, STA assumes all clocks and all paths are synchronous by default, and it will attempt to close timing for all synchronous paths. Designers need to specify asynchronous clocks or paths, where timing closure is not required.

What does being "synchronous" mean?

You may wonder what "synchronous" really means in the context of timing verification. If the relationship between launch clock and capture clock is bounded, we call it "synchronous". Some exceptions of being "synchronous" would be:

1. Launch clock is 100MHz, while capture clock is 37MHz
2. Launch clock and capture clocks do not share the same clock source, thus they don't have different skew or jitter
3. Launch clock tree and capture clock tree are not balanced

Conclusion

Understanding the nature of STA establishes the baseline of STA. We highly recommend readers to fully digest these concepts.

Q3: What are setup time and hold time?

The setup time is the amount of time needed by a register to have its input value stable **before** its triggering clock edge, or the amount of time needed by a latch to have its input value stable **before** the end of its transparent phase.

On the contrary, the hold time is the amount of time needed by a register to have its input value stable **after** its triggering clock edge, or the amount of time needed by a latch to have its input value stable **after** the end of its transparent phase.

Q4: What are setup time and hold time constraints?

Setup Time Constraints

Let us use T_c to denote clock period, t_{pcq} to denote clock-to-q propagation delay, t_{pd} to denote propagation delay of the combination logic between launch flop and capture flop, and t_{setup} to denote capture flop setup time.

The setup time constraints can be represented by:

$$T_c \geq t_{pcq} + t_{pd} + t_{setup}$$

It is easy to observe that, the propagation delay between launch flop and capture flop must satisfy:

$$t_{pd} \leq T_c - (t_{pcq} + t_{setup})$$

Designers must carefully balance and pipeline the design, such that t_{pd} does not exceed its upper bound.

Hold Time Constraints

Let us use t_{ccq} to denote clock-to-q contamination delay or the minimum clock-to-q delay, use t_{cd} to denote contamination delay or minimum delay of the combination path from launch flop to capture flop, and use t_{hold} to denote capture flop hold time.

The hold time constraints can be represented by:

$$t_{ccq} + t_{cd} \geq t_{hold}$$

It is easy to observe that, the minimum delay of the combination path has a lower bound:

$$t_{cd} \geq t_{hold} - t_{ccq}$$

There is one fundamental difference between setup time and hold time constraint: the setup time constraint is clock frequency dependent, while the hold time constraint is not. This is important, as this guides us the fix order during timing ECO.

Taking Clock Skew Into Account

In real silicon, the absolute time that clock rising edge arrives at launch flop and capture flop are different, and this introduces clock skews. In a more accurate timing modeling, clock skew has to be taken into account.

Clock skews can be positive when the clock source is closer to the launch flop, and both "data" and "clock" travel in the same direction. See the diagram below:

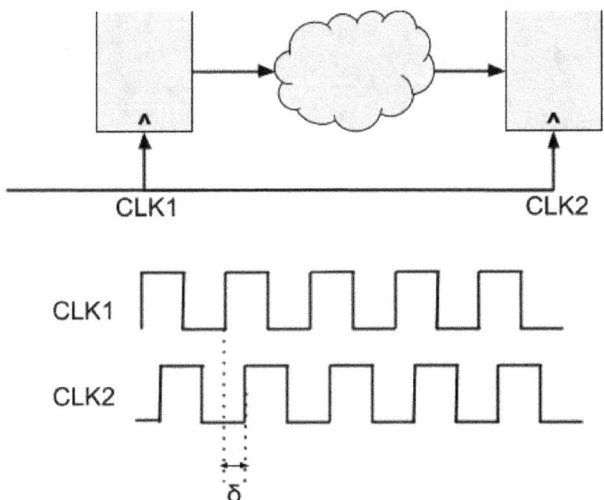

Obviously, positive clock skew helps setup timing closure, but hurts hold timing closure.

Clock skews can be negative when the clock source is closer to capture flop, and "data" travels in the opposite direction of "clock". See the diagram below:

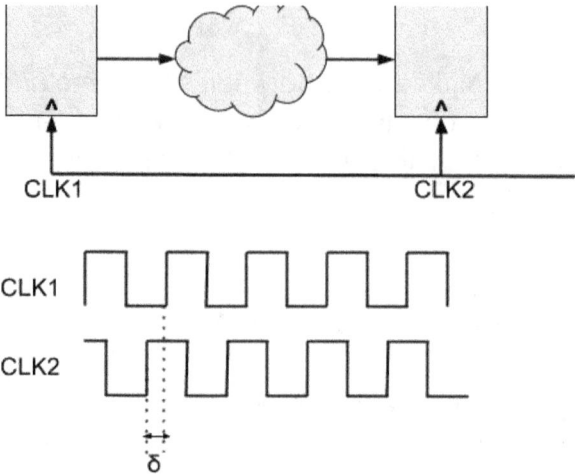

Obviously, negative clock skew helps hold timing closure, but hurts setup timing closure.

Now, the revised setup time and hold time constraints are shown below:

$$T_c + \delta \geq t_{pcq} + t_{pd} + t_{setup}$$

And:

$$t_{ccq} + t_{cd} \geq t_{hold} + \delta$$

Where δ denotes clock skews.

Conclusion

The setup time and hold time constraints set the foundation of STA, and these constraints should be kept in mind at all times.

Reference

Digital Integrated Circuits: A Design Perspective by Jan M. Rabaey, Anantha Chandrakasan, Borivoje Nikolic

Q5: What is the benefit of using half-cycle-path?

Until now, we focused on the timing of one-cycle-path or full-cycle-path. Sometimes, there may exist half-cycle-paths in design. One example will be the launch flop is negedge triggered, while the capture flop is still posedge triggered.

We discussed the clock skew and how it affects STA. In this example, half-cycle-paths can be modeled as one-cycle-paths with clock skew $\delta = -T/2$. It is obvious that hold time closure is easier while setup time closure is harder for half-cycle-path.

You may wonder what is the benefit of using half-cycle-path in the design.

Assuming the transmitter wants to transfer 32B data, and the clock is forwarded along with data (or the data is source synchronized). Such a path will have a large positive clock skew δ, thus setup time closure will be straightforward. However, hold time checks may have violations.

To fix hold time in such case, there are 2 possible solutions:

1. Add buffers in datapath, i.e., trading the setup time for hold time
2. Make launch flop negedge triggered while capture flop posedge triggered, i.e., using half-cycle-path

Apparently, the 2nd solution is cleaner.

As a side note, since we forward multiple bits with the clock, data-to-data skew needs to be checked by STA and balanced in physical implementation.

Q6: What are the sources for clock uncertainty?

You may already have noticed that, in timing reports, there is an item called clock uncertainty. Clock uncertainty hurts both setup and hold timing closure.

Sources of Clock Uncertainty

The exact amount of clock uncertainty usually cannot be predicted beforehand. The sources of clock uncertainty include:

1. Clock sources, e.g. PLL, have jitters in nature.
2. Manufacturing device variations
3. Changes of temperature during operation. The variation of temperature leads to variations of clock cell speeds
4. Power supply variations. The power rail will not always stay in the nominal voltage. For example, when more circuits are turned on and absorbing current, the voltage drop in the supply rail will be higher
5. Changes of capacitive load. Capacitive load is nonlinear in nature and its value depends on the applied voltage. In addition, for latches and flops, their clock load is a function of the stored value
6. Capacitive and cross-talk to adjacent wires

Modeling Clock Uncertainty

Usually, STA tools require designers to specify the clock uncertainty as a percentage of clock period. In ZWL synthesis or sanity synthesis, this value is more pessimistic. However, in timing verification of PnR netlist, designers use less pessimistic clock uncertainty but rely on On Chip Variation (OCV) or Advanced On Chip Variation (AOCV) to leave more timing margins in design.

Reference

Digital Integrated Circuits: A Design Perspective by Jan M. Rabaey, Anantha Chandrakasan, Borivoje Nikolic

Q7: How does STA check reset removal & recovery / clock gating cell / data to data timing?

Other than setup and hold time constraints, there are other types of timing constraints that STA should check. These checks include but not limited to:

- Reset removal time and recovery time
- Clock gating cell timing
- Data to data timing

Reset Removal Time and Recovery Time

Usually an asynchronous reset has to be synchronized to achieve synchronous deassertion. Reset deassertion cannot be too close to valid clock edges, otherwise it may cause metastability. This constraint is called reset removal time and recovery time, which is shown in the diagram below:

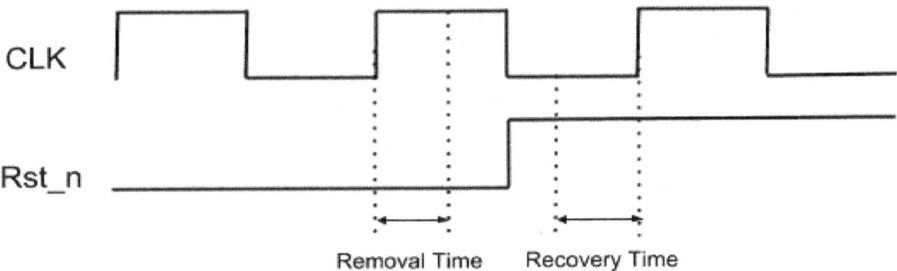

Reset deassertion must happen after reset removal time is met, but before entering the recovery time window. If you recall setup time and hold time, conceptually, removal time resembles hold time while recovery time resembles setup time.

STA should check all flops to satisfy reset removal time and recovery time constraints.

Clock Gating Cell Timing

Clock gating cells are widely used in SoC design, as a power-saving method. A typical clock gating cell, shown in the diagram below, consists of a low transparent latch and an AND gate. The existence of the latch is to make sure the clock does not glitch when the clock is high, but it also introduces additional timing checks.

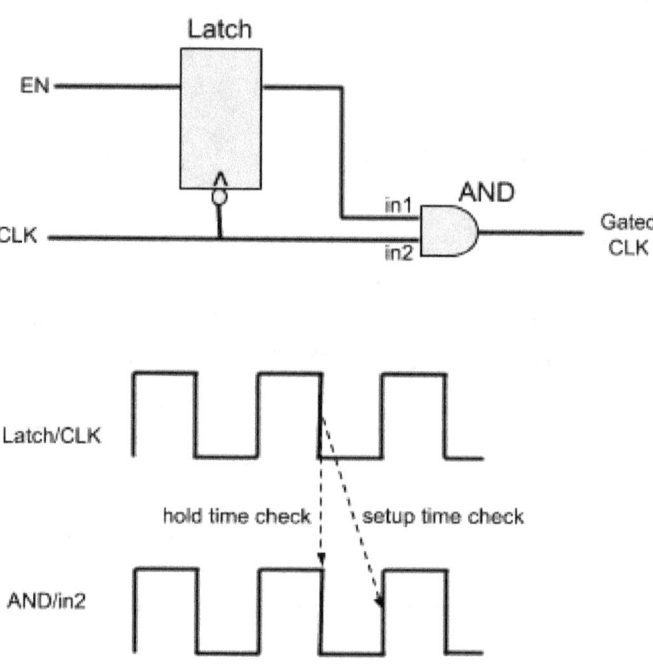

STA needs to make sure the "in1" becomes stable before the next positive clock edge, or setup time check has to be satisfied. STA also needs to guarantee that latch output change reaches AND gate after "in2" is low, or hold time constraint has to be met.

Data to Data Timing

Data to data timing check is to make sure data is stable relative to data valid. See the diagram below:

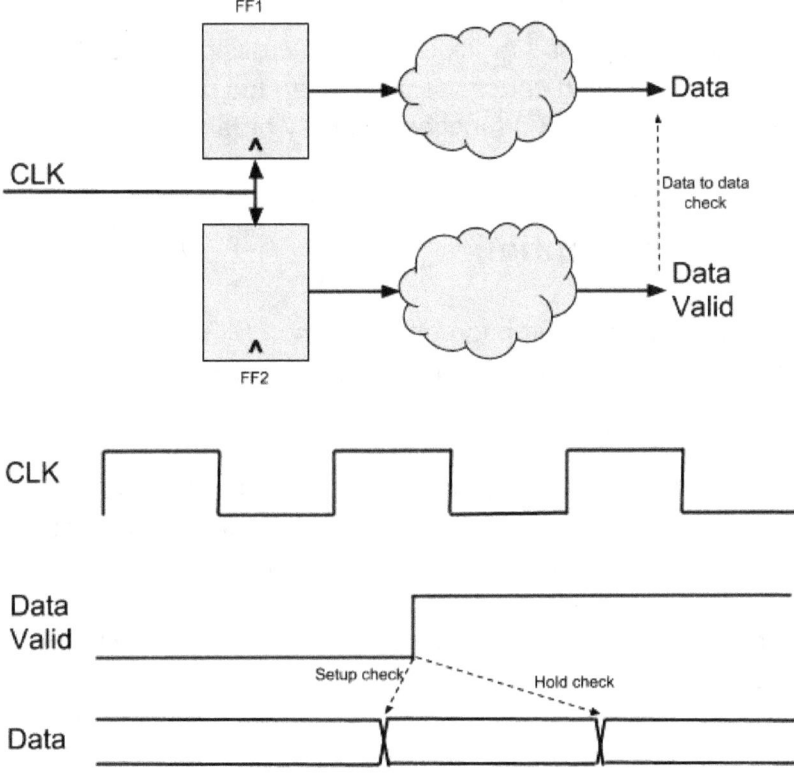

To be more specific, setup check is to make sure that data is stable before data valid asserts; hold check is to make sure that data is stable after data valid asserts.

Conclusion

We covered three special timing checks performed by STA. These timing checks are often used to check interviewee's understanding toward STA. In particular, interviewees should memorize the launch and capture clock edges with respect to each timing check.

Q8: How does STA verify async FIFO functionality?

Async FIFOs are widely used for clock domain crossing. Since STA is not used for async FIFO timing check, designers tend to naively ignore async FIFOs in STA and async FIFO functionality may break.

Async FIFO Assumption

There is a fundamental assumption about async FIFO: async FIFO pointers in source clock domain can only change 1 bit before synchronized to destination clock domain

Without special constraints, PnR tools may place different bits of the pointer far far away, thus multiple bits of pointers can change simultaneously from the synchronizers' point of view in the destination clock domain.

Therefore, designers need to constrain skews among different bits of the FIFO pointer.

How to Constrain Skews of Pointer Bits

If the pointers get updated every cycle, then the skew between bits must be within a bit time or 1 source clock cycle; if the pointers get updated every other cycle, then this constraint can be relaxed to 2 bit time or 2 source clock cycles.

Conclusion

We recommend readers to refer to Paul Zimmer's paper "No Man's Land, Constraining Async Clock Domain Crossings" for more implementation details. Understanding how to constrain asynchronous FIFO pointers will definitely impress your potential employers.

Reference

No Man's Land, Constraining Async Clock Domain Crossings by Paul Zimmer

Q9: How does STA check latch based design?

Why Do We Use Latch Based Design?

Latch based design enables time borrowing, which is shown below.

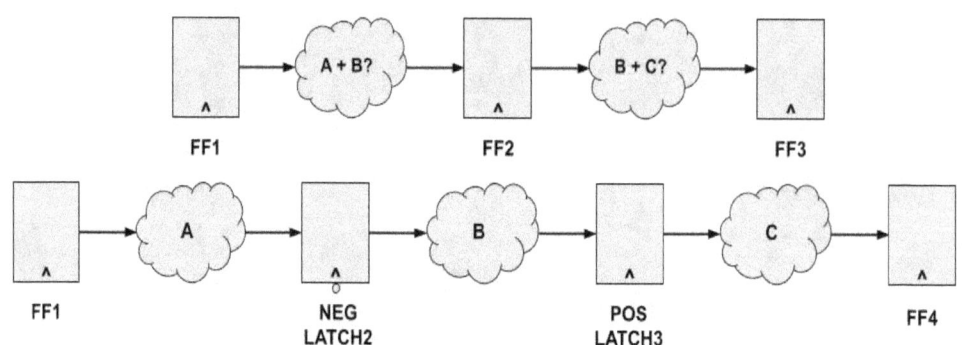

Say we want to finish computing logic A, B and C in 2 clock cycles, and none of these logic can be further divided. In pure flops based design, if both (A + B) and (B + C) take more than 1 cycle to finish, then it is impossible to close timing.

In latch based design, NEG LATCH separates logic A and B, and POS LATCH separates logic B and C. Now (A + B) can be relaxed to 1.5 cycles, and timing closure will be easier.

However, latch based design complicates the timing analysis.

How to Check Timing in Latch Based Design?

To check timing for logic A, see the diagram below. If logic A is able to finish in half cycle, then no time borrowing happens (Path I in the diagram); if logic A takes more than half cycle to complete, then there will be time borrowing (Path II in diagram).

In either case, logic A has to meet setup time constraint for NEG LATCH2.

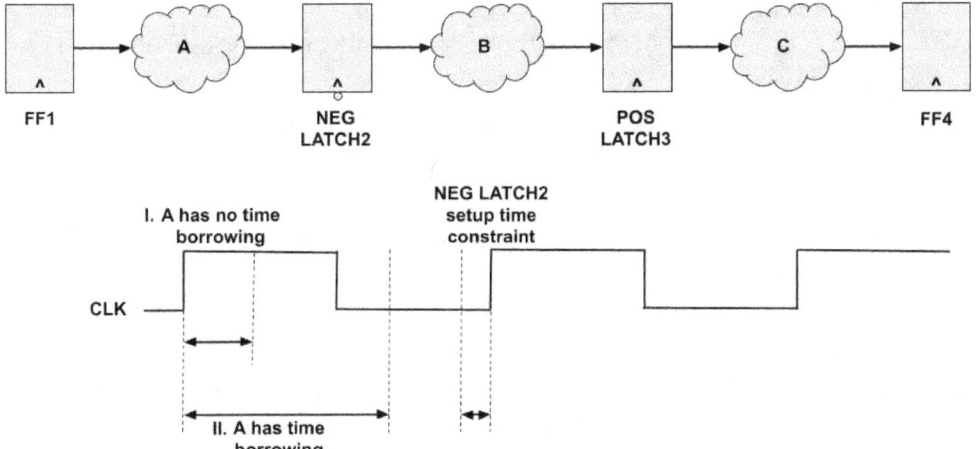

Checking timing for logic B becomes tricker, due to possible time borrowing in both logic A and B. See diagram below.

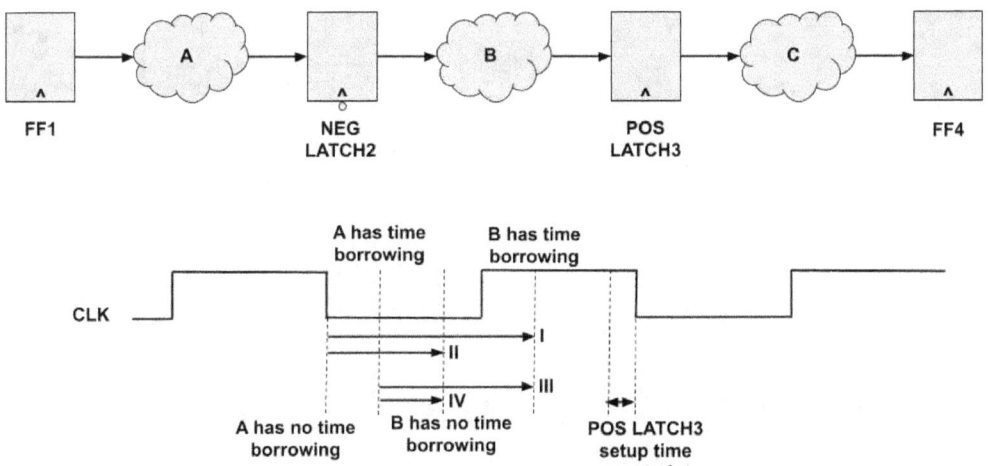

There are 4 cases to analyze:

1. Logic A does not have time borrowing, but B has time borrowing (Path I in diagram)

2. Logic A does not have time borrowing, and B does not have time borrowing either (Path II in diagram)
3. Logic A has time borrowing, but B has time borrowing (Path III in diagram)
4. Logic A has time borrowing, and B does not have time borrowing either (Path IV in diagram)

In any of the cases above, logic A and B have to meet setup time constraint for POS LATCH3

Logic C timing check has 2 cases as well, see the diagram below. Path I in the diagram shows the case where B has time borrowing, and Path II in diagram shows the case where B does not have time borrowing.

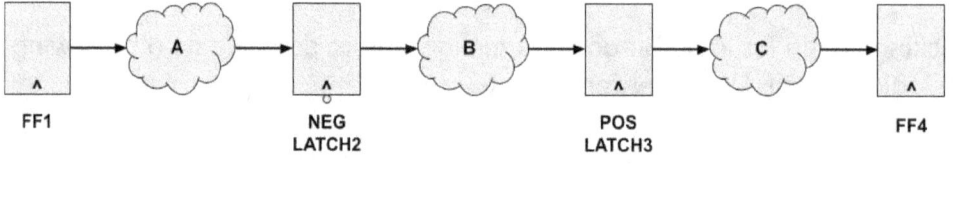

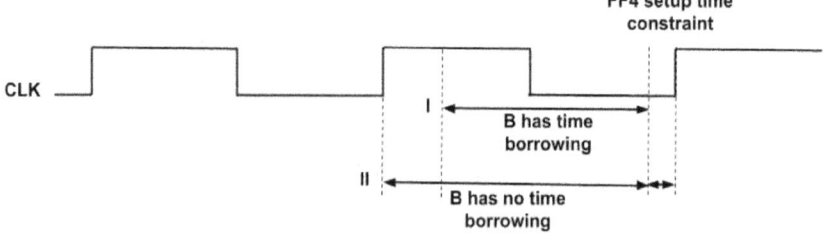

In either case, logic B and C have to meet setup time constraints for FF4.

Conclusion

Flop based design has a very clear boundary where the timing check should be performed, because flops are edge triggered. Latch based design has more cases for timing check, since latches are level sensitive in nature, and it enables time borrowing between different stages.

We recommend interviewees to try and explain latch based design timing checks in their own language.

Q10. How does multi-cycle-path (MCP) work in STA?

For one-cycle-path or full-cycle-path, Arrow 1 in the diagram below shows the default hold time checking edge, and Arrow 2 in the diagram below shows the default setup time checking edge.

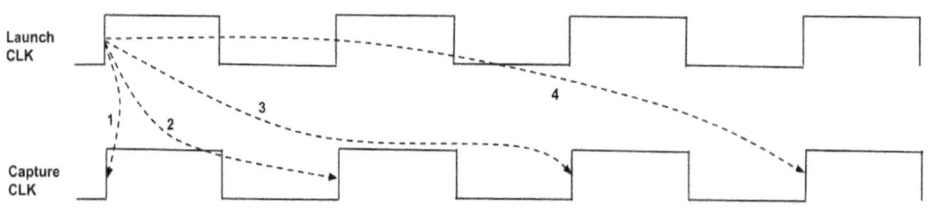

It is not uncommon to have multi-cycle-path in modern design. Multi-cycle paths can be used to constrain slowly changing signals, or quasi-static signals.

For a multi-cycle path of 3, Arrow 3 in the diagram above shows the default hold time checking edge, and Arrow 4 in the diagram above shows the default setup time checking edge.

A multi-cycle path of N can also be represented by the formulas below:

Setup time: $N * T_c \geq t_{pcq} + t_{pd} + t_{setup}$

Hold time: $t_{ccq} + t_{cd} \geq t_{hold} + (N - 1) * T_c$

It is easy to see that, with multi-cycle paths, setup timing is much easier to close, but hold time closure will be difficult.

In practice, designers may want to relax the hold time constraints by moving the hold time check to previous clock edges, depending on this multi-cycle path's functionality.

Chapter 2 Design Constraints (SDC)

Q11: What are design / library objects? How to access & manipulate these objects?

We have covered most of the Design Constraints (SDC) in our previous book "Verification, Implementation, Synthesis & Power" in the same book series. This chapter is intended to provide a quick recap, and also flag out a few Design Constraint differences between synthesis and physical design.

Design constraints are used throughout the implementation flow and physical design. They specify timing, power and area constraints for a design. Synopsys Design Constraints, or SDC, is a commonly used tcl based design constraint format. We will again use SDC as a case study in this book.

What Are Design / Library Objects?

These objects include:

1. **Design**: defined by top level module
2. **Port**: input, output and inout ports in the top level design
3. **Clock**
4. **Cell**: submodule instances in the design
5. **Pin**: submodule instance ports
6. **Net**: wire or registers to connect pins
7. **Library name**
8. **Library cell**

We can use the following commands to retrieve these objects:

```
# get design
get_design

# get ports
get_ports
all_inputs
all_outputs
```

 # get clocks
get_clocks
all_clocks

 # get cell
get_cells
all_registers

 # get pins
get_pins U1/A

 # get library name
get_libs16nm

 # get library cells
get_lib_cells 16nm/AND*

How To Access & Manipulate These Objects?

The following example shows how to access collections of objects:

 # get a list of all inputs
set input_list [all_inputs]

 # check number of inputs in the design
sizeof_collection $input_list

 # print out all inputs
query_objects $input_list

The following example shows how to add to and remove from the collections:

 # add all outputs to $input_list
set input_list [add_to_collection $input_list [all_outputs]]

```
# remove all outputs from $input_list
set input_list [remove_from_collection $input_list [all_outputs]]
```

The following example shows how to filter collections:

```
# get only AND* gates in the collections
# "ref_name" is an attribute of the object, used as filtering criteria
filter_collection [get_cells *] "ref_name =~ AND*"

# another way to get filtered collections
set and_gate_list [get_cells * -filter "ref_name=~ AND*"]
```

Note, other than "=~", there are other relational operators: "==", "!=", ">", "<", ">=", "<=", and "!~".

The following example shows how to iterate through collections:

```
# use "foreach_in_collection" to iterate over a collection
foreach_in_collection and_gate [get_cells * -filter "ref_name=~ AND*"] {
    echo "Instance [get_object_name $and_gate] is an AND gate."
}
```

Conclusions

Top-tier companies favor candidates with strong coding and scripting backgrounds. Getting familiar with these basic commands helps interviewees succeed in coding questions.

Q12: How to set single-clock design constraints in Post-CTS run?

During physical design, clock trees are synthesized, and clock paths are routed. Therefore, setting single-cycle design constraints in Post-CTS (Post-Clock-Tree-Synthesis) will be different from synthesis run. Interviewees should not mix synthesis run clock constraints and Post-CTS clock constraints.

First, we define the clock and its associated attributes, including clock period, waveform, name, and clock ports. If the clock duty cycle is not 50%, and both negedge and posedge are used in the design, then defining clock waveform is critical. This step is the same between synthesis run and Post-CTS run.

> *# Waveform matters if both negedge and posedge of the clock are used in the design*
> *create_clock -period 2 -waveform {0 0.5} -name CLK [get_ports clk]*

Next step is to model the clock uncertainty, which includes clock skew, jitter and margins. This clock uncertainty is usually set less pessimistically compared to synthesis run, as chip physical implementation details are available. There is no need to model clock skew anymore, and clock skew is extracted from the actual PnR netlist.

> *# Model clock jitters + margin; note no need to model skew*
> *set_clock_uncertainty -setup Tu [get_clocks CLK]*

Then we want to model the clock latency. Unlike synthesis run, this clock latency only includes the clock source latency, as clock network latency will be extracted from the PnR netlist.

> *# Model latency or insertion delay; note no need to model network latency in post layout*
> *# Source latency models the delay from the actual clock origin to the create_clock port or pin*
> *set_clock_latency -source -max 3 [get_clocks CLK]*

We use the "*set_propagated_clock*" command to calculate clock network latencies inside the current design. We recommend propagating the clock with "*[get_ports CLK]*" instead of "*[get_clocks CLK]*", to avoid propagating the I/O reference clock.

    # Actual propagated clock network latencies are calculated inside the current design,
    # while estimated ideal clock network latencies still apply to I/O reference clock
    set_propagated_clock [get_ports CLK]

Unlike synthesis run, we do not specify clock rise and fall transition time, since these are extracted and derived from the PnR netlist as well.

Q13: How to set I/O constraints for single-clock design in Post-CTS run?

Specifying only clock constraints is not enough for timing analysis. The STA tool needs to understand the timing information with each input or output port. Note, there are slight differences about how to set I/O constraints in synthesis runs and Post-CTU runs.

How to Specify Input / Output Delay?

The following example shows how to specify input port delay as if the input port is driven by a register.

```
# Constraining input paths; timing slack used by upstream logic
set_input_delay -max 0.6 -clock CLK [get_ports A]
# If the upstream logic is negedge triggered
set_input_delay -max 0.6 -clock CLK -clock_fall [get_ports A]
```

The following example shows how to specify output port delay as if the output port drives a register.

```
# Constraining output paths; timing budget left for downstream logic
# Output delay = Tsetup of capture FF + comb logic driven by the output port
set_output_delay -max 0.8 -clock CLK [get_ports B]
# If the downstream logic is negedge triggered
set_output_delay -max 0.8 -clock CLK -clock_fall [get_ports B]
```

Note, "*set_input_delay*" and "*set_output_delay*" will inherit the master clock's source latency in the delay calculation. By default, the delay value specified in "*set_input_delay*" and "*set_output_delay*" does not include clock latencies.

If we want to handle different I/O clock latencies from the master clock latency, we can include different clock latencies in "*set_input_delay*" and "*set_output_delay*", with "*-source_latency_included*" and "*-network latency_included*" options turned on.

Or, more commonly, we can create virtual clock copies of the master clock.

How to Specify Virtual Clock in Input / Output Delay?

The following example shows how to specify virtual clock I/O reference latencies. Note, we do not specify clock network latency for master clock in Post-CTS run.

create_clock -period 2 [get_ports clk]
 # launch virtual clock
create_clock -period 2 -name vclk_in
 # capture virtual clock
create_clock -period 2 -name vclk_out

 # specify clock source latency
set_clock_latency -source -max 0.3 [get_clocks clk]
set_clock_latency -source -max 0.18 [get_clocks vclk_in]
set_clock_latency -source -max 0.3 [get_clocks vclk_out]

 # specify clock network latency only for virtual clocks
~~*set_clock_latency -max 0.12 [get_clocks clk]*~~   *# No longer needed*
set_clock_latency -max 0.12 [get_clocks vclk_in]
set_clock_latency -max 0.07 [get_clocks vclk_out]

set_input_delay -max 0.6 -clock vclk_in [all_inputs]
set_output_delay -max 0.8 -clock vclk_out [all_outputs]

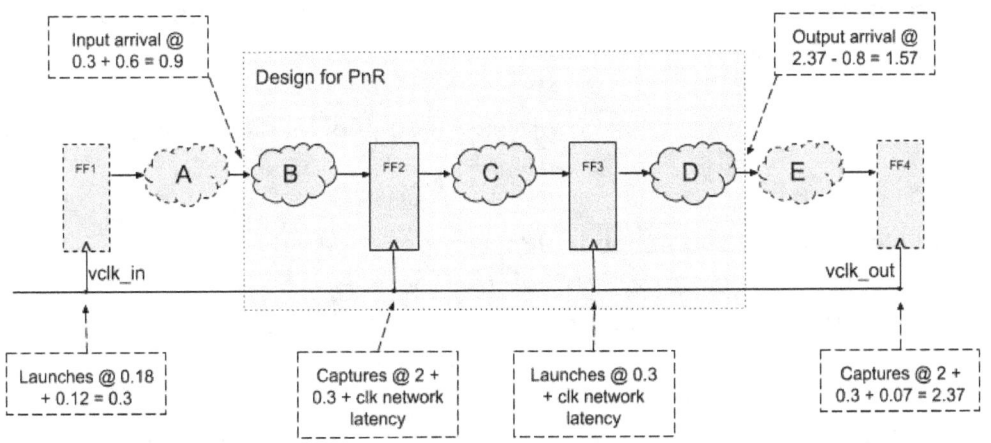

How to Avoid Propagating IO Reference Clocks?

We mentioned before that actual propagated clock network latencies are calculated inside the current design, and we shall use "*set_propagated_clock [get_ports CLK]*".

If we do "*set_propagated_clock [get_clocks CLK]*", then I/O reference clocks also become propagated, but there are no actual clock paths, thus calculated I/O clock latencies are 0.

To avoid this, we can always do "*set_propagated_clock*" with "*get_ports*"; or if we have to do "*set_propagated_clock*" with "*all_clocks*", we can create virtual clock for input/output delays and specify the source/network latency value on it. STA tool will preserve clock latency values specified for virtual clocks during propagation.

How to Specify Output Loads and Input Transition Time?

"*set_input_delay*" and "*set_output_delay*" are required but not sufficient for accurate timing analysis and optimization of IO paths. There are additional constraints we need to specify. Remember a gate's delay depends on output load and input slew.

The following example shows how to specify output loads and input transition times.

```
# Need to take into account output loads
set_load -max [expr {30.0/1000}] [get_ports B]
# The following 2 commands help to specify the load
# when absolute load value is not available
set_load -max [load_of my_lib/AN2/A] [get_ports B]
set_load -max [expr {[load_of my_lib/inv1a0/A] * 3}] [get_ports B]

# Need to take into account input transition times
set_input_transition -max 0.12 [get_ports A]
# The following 2 commands help to specify the transition time
# when absolute transition time value is not available
set_driving_cell -max -lib_cell OR3B [get_ports A]
set_driving_cell -max -lib_cell FD1 -pin Qn [get_ports A]
```

Sometimes STA engineers do not specify output loads and input transition times when closing block level timing. This is because block level interface timing will be addressed in full-chip level STA.

Q14: How to set multi-synchronous-clock design constraints?

How to Constrain Multiple Clock Input Delay?

Let's assume, the design uses clock "CLKC" with period of 2ns, and the register driving the inputs of the design uses clock "CLKB" with period 3ns. Both "CLKC" and "CLKB" are divided from the same reference clock, thus they are synchronous.

Thus we can define "CLKC" as master clock and "CLKB" as virtual clock to the design, and set input delay accordingly.

> # define master clock "CLKC"
> create_clock -period 2 [get_ports CLKC]
> # define virtual clock "CLKB"
> create_clock -period 3 -name CLKB
> # set input delay for all inputs
> set_input_delay -max 0.55 -clock CLKB [all_inputs]

The diagram below shows how synthesis / STA tools try to close timing. There are 2 possible timing paths, and the tools will try to close timing with tighter constraints.

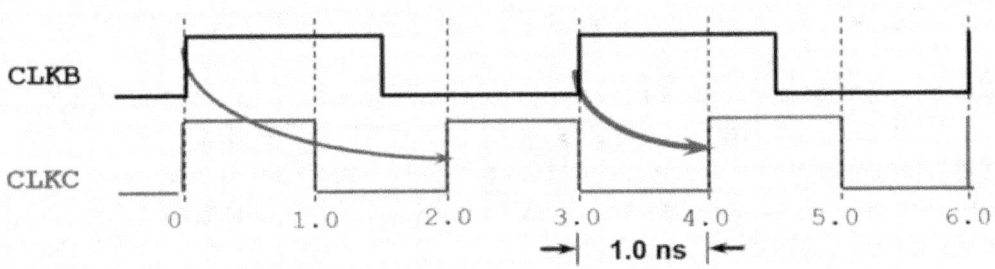

37

How to Constrain Multiple Clock Output Delay?

Assuming the design uses clock "CLKC" with period of 2ns, and outputs of the design drives registers clocked by either "CLKD" or "CLKE". The period of "CLKD" is 1.33ns, and the period of "CLKE" is 1ns. Let's also assume "CLKC", "CLKD" and "CLKE" are divided from the same reference clock, thus they are synchronous.

We can define "CLKC" as the master clock, and "CLKD" and "CLKE" as virtual clock to the design, and set output delay accordingly.

Note, without the "-add_delay" option, the 2nd "set_output_delay" will override the 1st one. This is not the desired behavior: we want tools to consider both clocks and meet timing for both constraints.

```
# define master clock "CLKC"
create_clock -period 2 [get_ports CLKC]

# define virtual clock "CLKD" and "CLKE"
create_clock -period 1.33 -name CLKD
create_clock -period 1 -name CLKE

# set output delay for all outputs
# -add_delay option must be specified to avoid override
set_output_delay -max 0.1 -clock CLKD [all_outputs]
set_output_delay -max 0.2 -clock CLKE -add_delay [all_outputs]
```

The diagram below shows how synthesis / STA tools try to close timing for the constraints above. Timing closure between "CLKC" and "CLKE" is relatively easy, since there is only one possible timing paths. There are 2 possible timing paths between "CLKC" and "CLKD", and the tool will try to close timing with tighter constraints.

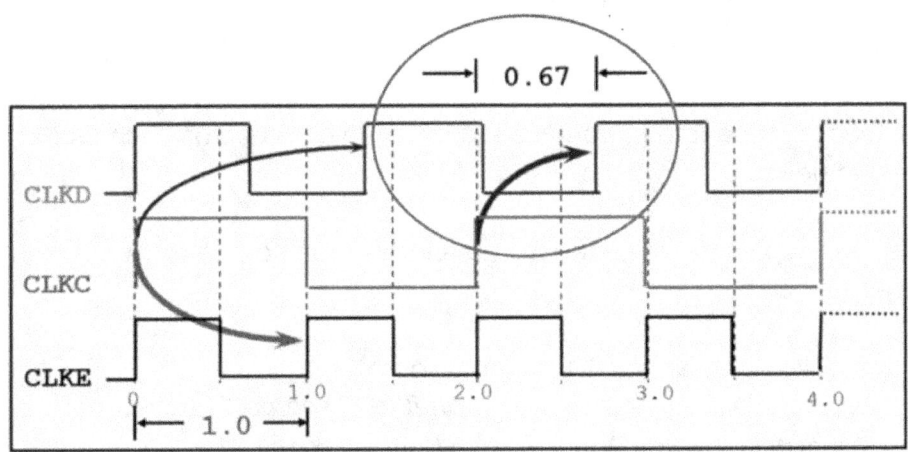

Q15: How to set generated clock design constraints in Post-CTS run?

For clocks propagated through sequential logic or macros such as PLL, we need to define generated clocks.

The first step, is to define the master clock or the source clock of the generated clock:

> # define master clock or source clock
> create_clock -period 2 -waveform {0 0.5} -name CLK [get_ports clk]
> set_clock_uncertainty -setup Tu [get_clocks CLK]
> set_clock_latency -source -max 3 [get_clocks CLK]
> set_clock_transition -max Tt [get_clocks CLK]

The next step is to define the generated clock. Generated clock is defined in hierarchical pins inside the design.

> # define generated clock
> create_generated_clock -divide_by 2 -name CLK_GEN
>   -source [get_ports clk] [get_pins FF1/out]
> set_clock_uncertainty -setup Tu_gen [get_clocks CLK_GEN]
> set_clock_transition -max Tt_gen [get_clocks CLK_GEN]

Remember:

1. The source latency of the generated clock = master clock source latency + master clock network latency + sequential logic / macro internal latency
2. The network latency of the generated clock = estimated delay from FF1/out to register clock pins

Q16: How to set mutually exclusive synchronous clock design constraints?

If there exists mutually exclusive synchronous clocks in the design, designers have a few ways to specify the exclusivity of the clocks:

1. set_case_analysis
2. set_false_path -from [get_clocks CLK1] -to [get_clocks CLK2] & set_false_path -from [get_clocks CLK2] -to [get_clocks CLK1]
3. set_clock_groups -logically_exclusive
4. set_clock_groups -physically_exclusive

"*set_case_analysis*" is the most straightforward way, and it constrains which clock will propagate through. However, it introduces different timing modes, and increases total runtime.

For "*set_false_path*" or "exclusive" clock paths, DC will not optimize timing for them, and STA will not check timing for them either.

"*set_false_path*" is usually not preferred, since it can introduce undesired timing exceptions. Note "*set_false_path*" still impacts the SI or noise analysis.

"*set_clock_groups*" is the most recommended one, but there are some differences between "*-logically_exclusive*" and "*-physically_exclusive*". "*-physically exclusive*" does not consider the SI or noise effect, while "*-logically_exclusive*" does.

Let's consider a simple example below, where clock mux for "CLK1" and "CLK2" resides in the design, and these 2 clocks are mutually exclusive.

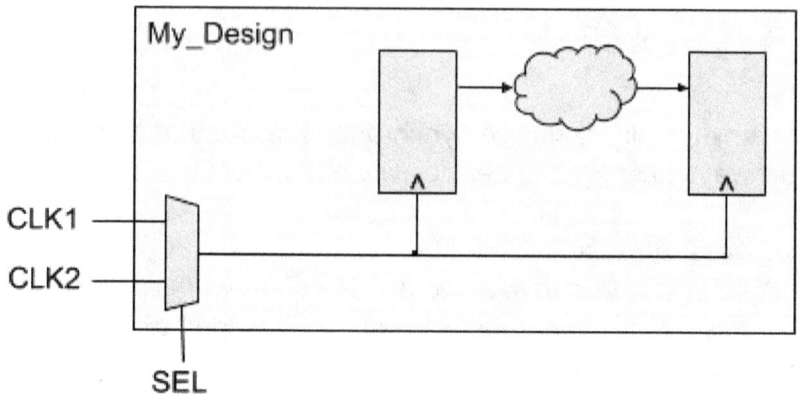

Since accurate crosstalk-induced delay analysis is required, we can use "*set_case_analysis*", "*set_false_path*", or "*set_clock_groups - logically_exclusive*".

```
# solution 1: use timing modes to verify timing
# mode 1
set_case_analysis 1 [get_ports SEL]
# mode 2
set_case_analysis 0 [get_ports SEL]

# solution 2: use false path
set_false_path -from [get_clocks CLK1] -to [get_clocks CLK2]
set_false_path -from [get_clocks CLK2] -to [get_clocks CLK1]

# solution 3: use logically exclusive
set_clock_groups -logically_exclusie -group CLK1 -group CLK2
```

What if the clock mux resides outside of the design? Then we do not consider SI effect anymore, thus physically exclusive shall be used.

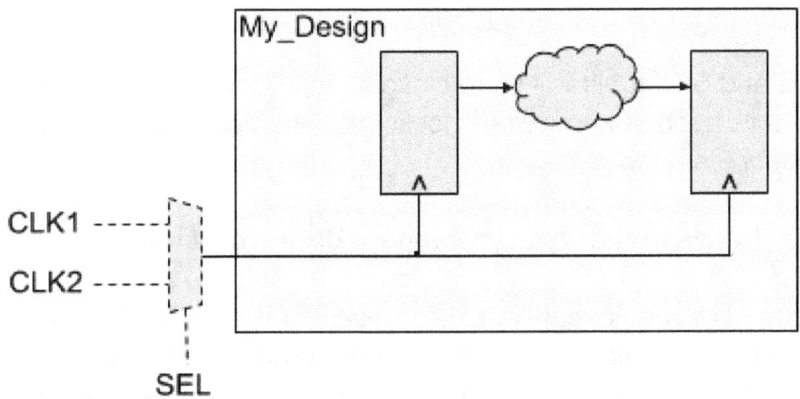

 # use physically exclusive
set_clock_groups -physically_exclusive -group CLK1 -group CLK2

In large designs, there may be hundreds or even thousands of clocks. Specifying "*set_clock_groups*" for mutually exclusive clocks is a cleaner and easier way to maintain. If using "*set_fase_path*", each exclusive clock pair will need 2 constraints, and the number of constraints will grow exponentially!

Q17: How to set asynchronous clock design constraints?

Synthesis and STA tools will try to close timing between synchronous clocks, thus the tools need to know what clocks or paths are asynchronous, thus no timing closure is needed.

There are 2 ways to specify asynchronous design constraints.

The first one is to use exceptions, i.e., specify "false path". Although intuitive and convenient, this approach is not recommended. If the path is not defined correctly and precisely, it may lead to unintended "false path" where timing closure is actually required.

The preferred approach is to use the "*set_clock_groups -asynchronous*" command. In addition, this command still considers SI or noise analysis with an infinite timing window. Thus this command is a closer modeling of real silicon.

Q18: How to verify SDC?

We could rely on the "*check_timing*" command in STA tools like PrimeTime, to verify the completeness and correctness of SDC. However, as the number of timing corners / scenarios increases and the runtime becomes longer, the STA team cannot afford to waste time running PrimeTime using "incomplete" or "incorrect" SDC. We need to have a way to "verify" SDC before actually doing timing verification. One of the most common tools is FishTail.

The table below shows a few example items that FishTail will check against SDC. It could also serve as a Design Constraint checklist. Usually, the STA team wants to make sure FishTail verification passes before running timing verification.

Clock Verification	
Clock Definition	Un-clocked registers
Clock Definition	Clocks defined on hierarchical pins
Clock propagation	Reconvergent clocks
Clock propagation	Clock propagating through non-unate cells whose side-inputs are not set to a constant value
Clock propagation	False path in clock generation logic that prevents the STA tool from propagating the clock
Generated clocks	Generated clock has no master clock
Generated clocks	Generated clock blocks the propagation of other clocks
Generated clocks	Generated clock has incorrect edge specification

Clock groups	Incorrect logically or physically exclusive clock groups
Clock groups	Incorrect clock-to-clock exceptions
Clock groups	Missing clock groups
Clock-to-clock false path	Missing timing exceptions between clocks with non-integer clock periods
IO delays	Missing input / output delays
IO delays	Propagation of conflicting constant values to the same pin
Case analysis	Case analysis values specified on multiple combo pins cannot be simultaneously satisfied
False-Path Verification	
False path	Make sure non-clock-to-clock false paths between synchronous clocks are defined correctly for setup timing check
Multi-Cycle-Path (MCP) Verification	
MCP	Prove when a startpoint transitions, this change cannot propagate to the endpoint in less than number of clocks cycles specified by user
MCP	A setup multi-cycle-path shall accompany with a hold multi-cycle-path
MCP	A hold shift is less than setup shift - 1

Chapter 3 STA Tool / PrimeTime

Q19: What is PrimeTime flow? Can you write a simple PrimeTime STA script?

PrimeTime is one of the most popular commercial STA tools in the market. We will use PrimeTime as a case study.

The timing analysis flow in PrimeTime can be divided into the following stages:

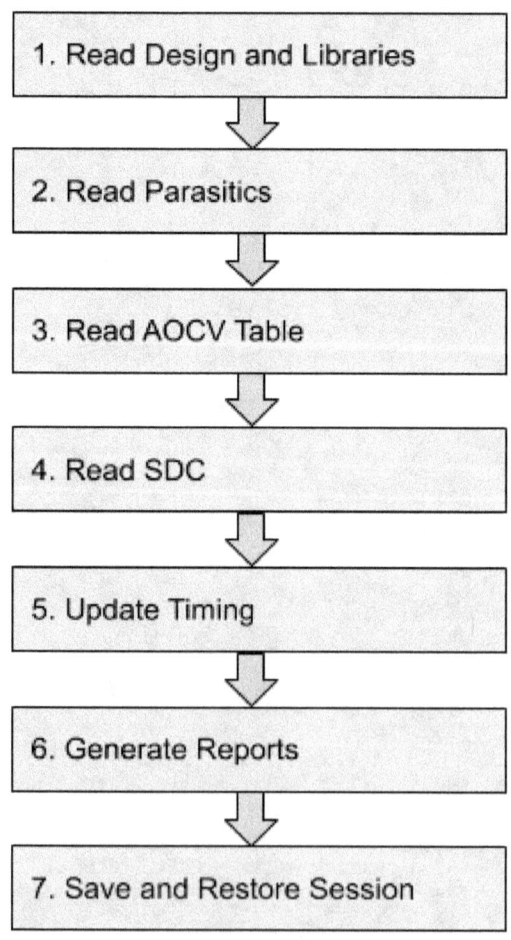

Read Design and Libraries

The following example shows how we read design and libraries in PrimeTime:

```
# read netlist
read_verilog my_design.v
# check current design
current_design

# link design and libraries
link_design
# check designs linked and report libs
list_designs
report_lib
```

Read Parasitics

Parasitics are RC information annotated to wires and chip internal interconnections, and PnR tool will generate parasitic values during place and route. Parasitics are typically in SPEF format, and are necessary to accurately model wire delay during timing verification.

The following example shows how we read parasitics in PrimeTime:

```
# read gzipped SPEF
read_parasitics top.spef.gz

# check parasitic annotation
# make sure all wires are properly annotated
report_annotated_parasitic -list_not_annotated
```

Read AOCV table

The following example shows how we read AOCV table in PrimeTime:

```
# read AOCV table
read_aocvm test.aocvm

# report AOCV annotation
report_aocvm -list_not_annotated
```

Read SDC

The following example shows how we read SDC in PrimeTime:

```
# read SDC
read_sdc top.sdc

# check completeness and correctness of SDC
check_timing -verbose
report_clock - skew -attribute
report_exceptions -ignored
report_case_analysis
```

Update Timing

This is the real meat of timing analysis. The following example shows how we invoke a full timing update in PrimeTime:

```
# invoke a full timing update
update_timing -full
```

Generate Reports

The following example shows how we should generate reports and check "*update_timing*" results in PrimeTime. Note, it is not recommended to use "*report_timing*" in the first step.

> *# generate summary reports*
> *report_global_timing [-pba]*
> *report_qor [-pba]*
> *report_analysis_coverage*
> *report_constraint -all_violators [-pba]*
> *report_disable_timing*
>
> *# generate detailed reports in GBA mode*
> *report_timing [-input_pins]*
>   *[-path full_clock_expanded]*
>   *[-net]*
>   *[-delay min/max]*
> *# generate detailed reports in PBA mode*
> *report_timing -pba_mode path*

Save and Restore Session

Before we exit, it is always a good practice to save the PrimeTime session, so next time we do not have to start from scratch. The following commands show how we save and restore PrimeTime sessions.

> *# save session*
> *save_session $rundir*
>
> *# restore session*
> *restore_session $rundir*

Conclusion

We showed a general PrimeTime flow and relevant commands. It will be a strong advantage if candidates can explain the PrimeTime flow in their own language.

Q20: What to check before running PrimeTime?

There are several things we need to check before actually running "*update_timing*":

1. Design should be fully read in by PrimeTime, and there shall be no black boxes or missing libraries
2. The netlist shall be accurately and completely annotated by SPEF, and there shall be no unannotated parasitics
3. The netlist shall be accurately and completely annotated by AOCV table; if there exists unannotated cell, we should debug why
4. The SDC needs to be read in correctly by PrimeTime, and there shall be no "unrecognized" SDC identified by PrimeTime
5. The SDC needs to be complete, for example, there is no missing clock, no un-clocked registers, no unconstrained ports, etc.

Designers should check all listed items above, in order to achieve accurate STA results. To check #4 and #5 above, design engineers can leverage commercial SDC verification tools like FishTail.

In some companies, such checks are performed automatically, and any error or warning will be captured in a dashboard.

Q21: What are graph based analysis (GBA) and path based analysis (PBA)?

You may have already noticed that we specified "*-pba_mode*" in the "*report_timing*" command. This relates to PrimeTime's different modes of analysis:

> Graph based analysis, or GBA
> Path based analysis, or PBA

GBA Mode

GBA mode is the default analysis in PrimeTime. It uses the worst input slew of a cell to calculate that cell's output transition. The calculated output slew will be used to calculate the delays of downstream cells.

GBA mode is pessimistic in the analysis of certain paths. If a path can meet timing in GBA mode, then there is no need to run PBA mode for this path.

PBA Mode

PBA mode performs path-specific slew propagation. It first computes path-specific cell and net delays, then computes path specific slew degradation across nets, and finally re-computes resulting path delay.

PBA mode is computationally prohibitive to apply to the entire design. It can only be applied on user specified timing paths.

PBA mode has two options: **path** and **exhaustive**.

PBA Mode: Path and Exhaustive

Path option is fast, but it does not guarantee that the reported paths are truly the ones with the worst slack. This is useful to get an idea how much improvement the PBA may achieve.

Exhaustive option is a global recalculation. It calculates up to 25000 paths per endpoint. It ensures that the reported paths are truly the ones with the worst recalculated slacks.

Conclusion

In the early stage of the development, runtime is the major concern, thus designers should consider first using GBA mode; at a more advanced stage or sign-off stage, accuracy is more important, thus designers should consider using PBA mode.

The differences between GBA mode & PBA mode, Path & Exhaustive are quite popular in hardware interviews, especially when it comes to checking interviewee's understanding of PrimeTime.

Q22: What are OCV / AOCV / POCV?

In manufacture, chips on the same die may suffer from variations due to process, voltage or temperature changes, thus the same transistor can be faster or slower in different dies. To compensate for the variation, STA introduces a concept called On Chip Variation, or OCV. During design time, extra timing margins are added in timing analysis.

OCV has been evolved to **Advanced OCV / AOCV**, or even **Parametric OCV / POCV**.

OCV

In OCV, all cells or nets in the launch clock path, data path and capture clock path will be added a fixed derate value, bringing more pessimism in timing analysis and compensating the variation.

AOCV

In the real world, variations are rarely constants. Blindly adding a fixed derate value not only makes it hard for timing convergence, but also introduces unnecessary area and power increase during timing closure.

Variations usually follow Gaussian distribution: the more levels of logic a path has, the closer its variation follows Gaussian distribution, and the less variation the path has.

AOCV is represented by a 2-dimensional table: the derate value of a cell is determined by logic depth and distance. We have already shown how to read the AOCV table in the PrimeTime flow, by using the "*read_aocvm*" command.

POCV

POCV models a cell delay using Gaussian distribution directly, instead of adding a derate value. Cell delay is calculated from a "parameter", which is extracted from either library, or POCV table.

Q23: How to calculate timing slack using OCV?

We will use an example to illustrate how to calculate timing slack using OCV.

An Example

Let's assume OCV will "derate" 10% for clock cells, and 20% for data paths. A reg-to-reg path along with delay values is shown in the diagram below.

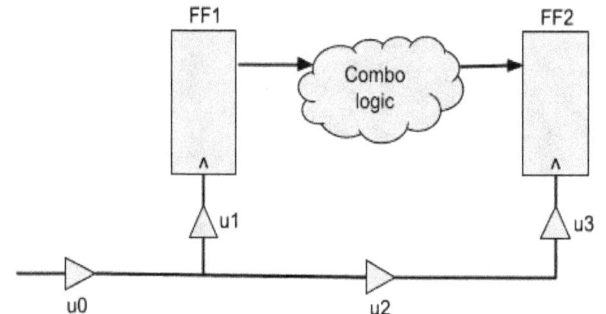

Delay values:

- $T = 1ns$
- $u0 = u1 = u2 = u3 = 0.1ns$
- FF1 clk-to-q delay = 0.2ns
- FF2 setup time = 0.1ns
- FF2 hold time = 0.1ns
- Combo logic Tmin = 0.2ns
- Combo logic Tmax = 0.4ns

Setup Timing Slack with OCV

Let's first consider the case without OCV.

> Launch clock path delay = 0.1 + 0.1 = 0.2ns
> Capture clock path delay = 0.1 + 0.1 + 0.1 = 0.3ns
> Clock skew δ = capture clock path delay − launch clock path delay
> = 0.1ns

Recall setup timing constraint:

$$t_{pcq} + t_{pd} + t_{setup} = 0.2 + 0.4 + 0.1$$
$$= 0.7ns < T_c + \delta = 1 + 0.1 = 1.1ns$$

Thus setup timing constraint is met, and the setup time slack is 1.1 - 0.7 = 0.4ns.

Taking OCV into account, for setup timing, both launch clock path and data path will be derated with a value larger than 1, capture clock path will be derated with a value less than 1. Therefore,

> Launch clock path delay = (0.1 + 0.1) x (1 + 10%) = 0.22ns
> Capture clock path delay = (0.1 + 0.1 + 0.1) x (1 - 10%) = 0.27ns
> Clock skew δ = capture clock path delay - launch clock path delay
> = 0.27 - 0.22 = 0.05ns

Observing that the clock skew δ becomes smaller, setup time closure will be harder.
Now the setup timing constraint becomes:

> $t_{pcq} + t_{pd} + t_{setup}$ = (0.2 + 0.4) x (1 + 20%) + 0.1
> = 0.82ns < $T_c + δ$ = 1 + 0.05 = 1.05ns

Thus setup timing constraint with OCV is still met, and the setup time slack is 1.05ns - 0.82ns = 0.23ns, which is also smaller than before. This means extra setup time margin is added after taking OCV into account.

Hold Timing Slack with OCV

Again, let's first consider our example without OCV.

> Launch clock path delay = 0.1 + 0.1 = 0.2ns
> Capture clock path delay = 0.1 + 0.1 + 0.1 = 0.3ns
> Clock skew δ = capture clock path delay - launch clock path delay
> = 0.1ns

Recall hold timing constraint:

> $t_{ccq} + t_{cd}$ = 0.2 + 0.2 = 0.4ns > $t_{hold} + δ$ = 0.1 + 0.1 = 0.2ns

Thus hold timing constraint is met, and the hold time slack is 0.4 - 0.2 = 0.2ns.

Taking OCV into account, for hold time, both launch clock and data path will be derated with a value less than 1, capture clock path will be derated with a value larger than 1. Therefore,

Launch clock path delay = (0.1 + 0.1) x (1 - 10%) = 0.18ns
Capture clock path delay = (0.1 + 0.1 + 0.1) x (1 + 10%) = 0.33ns
Clock skew δ = capture clock path delay - launch clock path delay
= 0.33 - 0.18 = 0.15ns

Observing that the clock skew δ gets larger, hold time closure may be harder.

Now the hold timing constraint becomes:

$t_{ccq} + t_{cd}$ = (0.2 + 0.2) x (1 - 20%)
= 0.32ns > t_{hold} + δ = 0.1 + 0.15 = 0.25ns

Thus the hold timing constraint with OCV still meets, and hold time slack is 0.32 - 0.25 = 0.07ns, which is also smaller than before. This means OCV adds extra hold time margin.

Conclusion

We used an example to explain how to calculate timing slack using OCV. The table below shows how we should "derate" the path delay in different scenarios. We recommend interviewees to fully digest the example and the table below.

	Launch Clock Path	Data Path	Capture Clock Path
Setup Time	Derate > 1	Derate > 1	Derate < 1
Hold Time	Derate < 1	Derate < 1	Derate > 1

Q24: What is CRPR in PrimeTime timing reports?

We used the example below to explain how derate / OCV works in PrimeTime. An interesting observation is that, in setup time and hold time analysis, cell u0 is subject to different derating. Obviously, this introduces unnecessary pessimism in timing analysis.

PrimeTime introduces a concept called Clock Reconvergence Pessimism Removal, or CRPR, to address this unnecessary pessimism.

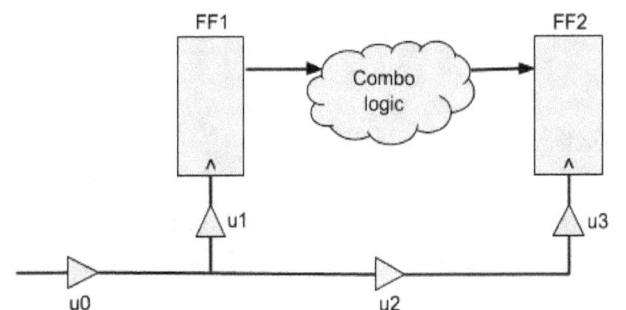

Delay values:
- T = 1ns
- u0 = u1 = u2 = u3 = 0.1ns
- FF1 clk-to-q delay = 0.2ns
- FF2 setup time = 0.1ns
- FF2 hold time = 0.1ns
- Combo logic Tmin = 0.2ns
- Combo logic Tmax = 0.4ns

A Formal Definition of CRPR

Clock Reconvergence Pessimism, or CRP, is the difference in delay along the common part of the launching and capturing clock paths. It assumes the shared segment has a min delay for one path and a max delay for the other.

This is an undesired effect due to the limitation of the STA tool. The removal of this pessimism is called CRPR.

A CRPR Example

Using the same example:

1. In setup timing analysis, the CRP for u0 = 0.1 x (1 + 10%) - 0.1 x (1 - 10%) = 0.02ns. By doing CRPR, setup time slack increases from 0.23ns to 0.25ns.
2. In hold timing analysis, the CRP for u0 = 0.1 x (1 + 10%) - 0.1 x (1 - 10%) = 0.02ns. By doing CRPR, hold time slack increases from 0.07ns to 0.09ns.

Actual CRPR in PrimeTime

In the example above, CRP for both setup time and hold time analysis is 0.02ns. However, for the same path, PrimeTime will not report the same CRP for setup and hold.

In fact, CRP in hold time is larger than that in setup time.

This is because hold time check is with respect to the same clock edge, while setup time check is with respect to 2 different clock edges. CPR in setup time has to count for the IR drop in the common path.

Conclusion

The concept of CRPR is often overlooked in interview preparation. We recommend interviewees to fully understand the concept of CRPR using your own example, and understand why CRP is different in setup time and hold time checks.

Chapter 4 Timing ECOs

Q25: What is the procedure for timing ECOs?

A Recap

Engineering Change Order (ECO) is the practice of introducing logic or gates directly into the netlist corresponding to a change that happens in RTL.

There are 2 types of ECOs: functional ECOs and timing ECOs.

Functional ECOs are necessary when RTL changes are required. This can happen due to RTL bug fixes, PPA fixes, or even last-minute spec changes.

Timing ECOs typically only touch PnR netlists. It is needed when, e.g., setup & hold timing violations are found during STA / timing verification.

Timing ECO Procedure

A typical timing ECO procedure is shown in the diagram below:

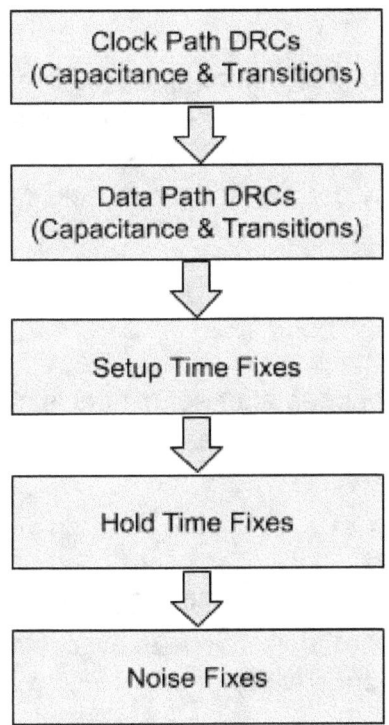

DRC Fixes

Design Rule Checks, or DRC, usually refer to max transition and max capacitance limits set either by the timing library, or by designers. DRC fixes are the first step for timing ECO, since this prevents any extrapolation while computing the cell delay from the timing library look-up table. If DRC violations are not addressed first, the delay values reported by STA tools are not accurate and not reliable.

Note, clock path DRC fix has higher priority over data path DRC fix.

Setup Time Fixes

Setup time fixes are performed before hold time fixes, because this allows more timing margin for hold time fixes. Remember, setup time constraints are clock cycle dependent, while hold time constraints are not. Some hold time fix techniques are essentially borrowing timing slacks from setup time.

Hold Time Fixes

Hold time fixes typically happen after setup time is clean.

Noise Fixes

Noise fixes shall be the last step of timing ECO since noise is very sensitive and designers cannot afford to fix noises at every single stage of timing ECO. Remember, any ECO change will impact the timing windows, thus the overall noise pictures.

Conclusion

The timing ECO procedure and related timing fix techniques are widely covered in hardware interview questions. Other than STA engineers and physical design engineers, this topic is often used to test RTL designers' understanding towards timing verification and implementation flow.

Q26: How to fix setup time and hold time violations?

Setup Time Fixes

There are several ways for setup time fix:

1. Vth swap
2. Cell upsizing
3. Sizing cells on side branch
4. Clock skew adjustment

The first three techniques are straightforward, and the last one is worth some discussion.

Clock skew adjustment means borrowing timing slacks from later stages or previous stages. Borrowing timing slack from later stage requires delaying the clock at the capture flop; borrowing timing slack from previous stage requires removing clock buffers of launch flop.

Usually there is less room to play with by borrowing timing slack from previous stages, because this is bounded by how many clock buffers in the launch clock path.

Hold Time Fixes

Some of hold time fixes are essentially opposite to setup fixes, for example:

1. Vth swap
2. Cell downsizing
3. Add dummy load or upsize cells on side branch

These techniques are "trading off" setup time over hold time slack.

There is one more technique that is specific to hold fix: buffer insertion. Designers need to be cautious about where to insert the buffers:

1. Inserting buffers at common point, results in optimal number of buffers to fix hold time; however, this may impact setup time closure for other unrelated timing paths
2. Inserting buffers at endpoint, results in more number of buffers being inserted; however, hold time fix rate will be more predictable and deterministic, as the impact on setup time would be minimal

Q27: What are timing ECO tools?

We have covered the typical timing ECO procedure, and common timing fix techniques. Nowadays, designers also heavily rely on timing ECO tools to do the job. We will cover 2 major timing ECO tools in this post:

    PrimeTime ECO
    Tweaker ECO

PrimeTime ECO

PrimeTime has an ECO flow for DRC or timing fixes, for example:

```
# DRC fixes
fix_eco_drc -type max_transition
fix_eco_drc -type max_capacitance
fix_eco_drc -type max_fanout

# report DRC fix results
report_constraint -all_violators -max_tran -max_cap -max_fanout

# timing fixes
fix_eco_timing -type setup
fix_eco_timing -type hold -buffer_list {buffer_lib_cell}

# report timing fix results
report_analysis_coverage
```

In addition, it can perform cell sizing and Vt swap using the same command:

```
# replace
size_cell {cell_list} {lib_list}
```

Tweaker ECO

Tweaker ECO was a platform developed by Dorado ECO company, targeting ECO solutions, and is now part of Synopsys EDA portfolio.

Tweaker ECO, in general, offers great correlation between ECO analysis and post PnR timing results. It takes the library, PnR netlist, SDC constraints, SPEF, AOCV / POCV table and DEF (floorplan) as inputs. Users can perform ECO experiments in its GUI, and see the timing results in real time. This helps to reduce both ECO time and number of STA iterations.

Conclusion

If interviewees have some basic understanding of these two timing ECO tools we mentioned here, it will definitely impress the potential employer. Readers can find more information on Synopsys website.

Part 2 Silicon Debug

Q1: What are common Design for Debug (DFD) techniques?

As the SoC complexity increases significantly, silicon debug becomes more and more difficult. Without special handling, identifying the root cause of chip failures would be nearly impossible. Therefore, designers will need to utilize several Design for Debug, or DFD techniques during chip design. These techniques offer great visibilities towards silicon debugging, and they are widely adopted in today's SoC design.

Exposing Internal States to Register Space

Exposing system internal states to register space is the most straightforward solution. These internal states include but are not limited to:

1. FIFO full / empty status
2. FIFO count
3. Credit counter status
4. FSM status

One thing to remember is that, these internal states shall not be dynamic changing, since register access is usually slow. Dynamic changing states captured by register access can be stale and they do not reflect the system real-time status. Therefore, this technique is more suitable for debugging a hang condition.

Security is another concern, since hackers can get system internal states by accessing registers. Usually some "locking" and "hiding" mechanisms shall be implemented to protect these debug registers being exposed in production chips.

Debug Counters and Performance Counters

Designers can implement debug counters in RTL, counting the number of occurrences of certain predefined error events. These debug counters are useful in silicon validation.

Designers often implement performance counters in RTL as well, counting the number of predefined performance events.

For example, in DDR memory controller, performance counters can be used to track the number of activation, read, write, pre-charge, refresh commands issued to DDR devices during a certain amount of time, and see if read and write commands are uniformly distributed across DDR banks and bank groups; later on, based on these performance statistics, silicon validation engineers and product engineers can further tune the DDR controller configuration knobs, to avoid traffic "hotspot" and improve controller performance.

Another example would be implementing "stall" counters in a valid-ready handshake interface. 3 sets of counters can be used to track the occurrences of the following events during a certain amount of time:

1. Out of reset, both valid and ready are low
2. Out of reset, valid is high but ready is low
3. Out of reset, valid is low but ready is high

If Event 2 happens often, it implies the receiver has a performance bottleneck, compared to the sender; if Event 3 happens often, it implies the sender has a performance bottleneck, compared to the receiver.

Interrupts

Interrupts are a computer system mechanism where IO devices, software applications, or the CPU asserts a signal or send a message to the CPU itself, some higher level software, or operating system that it needs attention or something needs service.

Interrupts can signal normal events like the device has finished a task, or problematic situations like TLB miss, page fault, and a branch condition misprediction in CPU out-of-order scheduling.

In RTL, design engineers need to implement some hardware-write-1-set and software-write-1-clear config register bits, and route these interrupt bits to CPUs or some special hardware called Programmable Interrupt Controllers (PIC). In addition, each interrupt bit typically pairs with one interrupt enable bit, and the interrupt is masked off if the interrupt enable bit is set to 0.

In silicon debug, interrupts can signal if a predefined error event occurs, and validation engineers can easily figure out the cause of the error event.

Scan Dump

DFT engineers will insert scan chains in the chip during synthesis. The scan chains are typically used to detect stuck-at faults, bridging faults and open faults.

The internal scan methodology works like this:

1. In the **scan shift-in phase**, scan data is shifted into the scan chain, one bit per clock. Scan shift-in phase can take multiple clock cycles to fill all the flops in a chain
2. In the **scan load phase**, all the scan flop outputs are applied to the inputs of the combinational logic between flop stages, and the combinational logic outputs are captured in flops which these combinational logics are driving. Load phase may happen in one clock cycle
3. In the **scan shift-out** phase, the scan flop outputs are shifted out from the scan chain for further analysis, one bit per cycle. Similar to scan shift-in phase, this phase can also take multiple clock cycles

The scan chains are also useful in silicon debug, especially in debugging system hang conditions. When a system hang happens, validation engineers can control the chip to operate under scan shift-out phase, and capture the value of every single flop in the system.

Validation engineers and design engineers can then work together to check, for example,

1. If the configuration registers are programmed properly
2. Which state the FSM is stuck at
3. If there exists credit leaks in a credit-based handshake interface

On-chip Trace Analyzer

On-chip trace analyzer is a special design to record a stream of predefined events.

An example of the on-chip trace analyzer is shown in the diagram below:

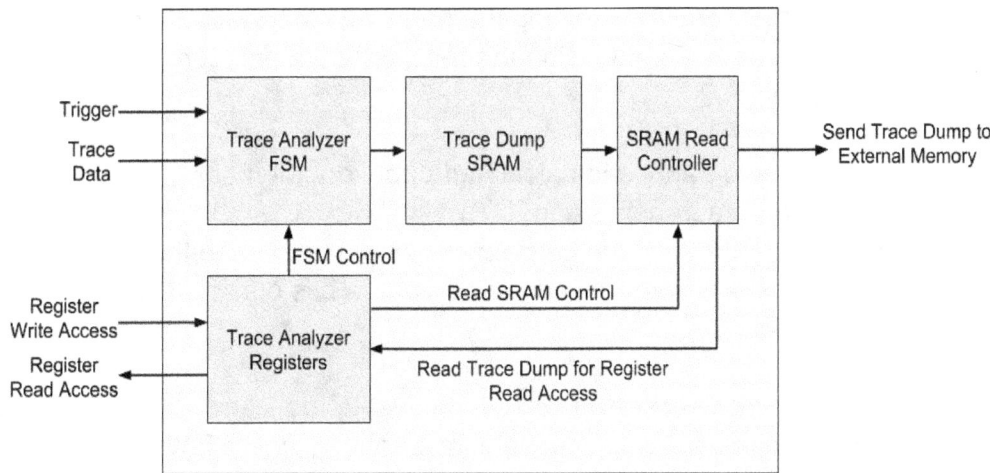

Usually, designers need to define some trigger events and trace data to dump. For example, a trigger event can be a "start" signal, and trace data can be the states of an FSM.

Once the trigger event happens, trace analyzer will start capturing trace data and store it along with some time stamp in the trace dump SRAM. The behavior of capturing trace data can be configured through registers.

Once the trace dump is stored in SRAM, it can either be read through register accesses, or sent to external memory. Storing trace dump to external memory like DRAM can be useful, since trace dump SRAM has limited storage.

Parity, ECC, Checksum & CRC

All techniques discussed earlier are suitable for control path debugging. For datapath debugging or data integrity checks, parity, ECC, checksum and CRC are all useful.

Parity is a simple and inexpensive way to detect errors during data transmission. One example would be generating 1-bit parity bit by XOR-ing all data bits together, and transmitting the parity bit along with the data. The receiver recalculates the parity using the received data and compares it with the received parity:

1. If the received parity and calculated parity do not match, then there is an error in data transmission
2. If the received parity and calculated parity match, then there is no error, or there are errors in even number of bits during data transmission

Error Detection and Correction (ECC), not just detects errors, but also fixes errors as well. It takes a similar mechanism as parity, where both ECC and data are transmitted together. At the receiver side, ECC is recalculated and compared with the received ECC. For example, DDR memory uses a form of Hamming code as ECC code, which is capable of detecting up to 2 error bits and correcting up to 1 error bit.

Checksum & Cyclic Redundancy Check (CRC) are the other two widely used data error detection methods. One fundamental difference is, checksum is not data order dependent, and it can work with out-of-order data streams, while CRC is data order dependent.

Assuming we are implementing some data integrity check in an IO device, and the device will issue read requests to the memory controller. Remember DDR memory may return read data out of order, thus the IO device may need a data reorder buffer before the data can be processed. See diagram below:

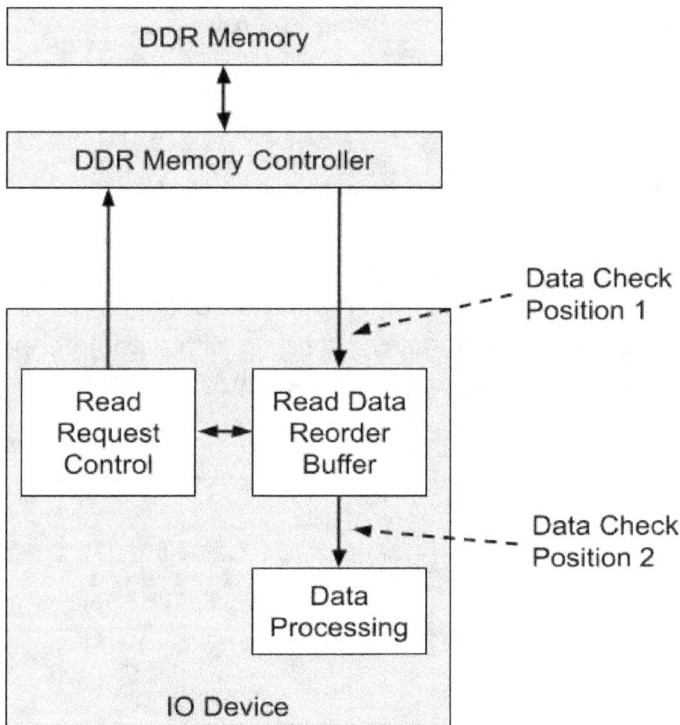

Checksum can be implemented at both Position 1 and 2 above, but CRC can only be implemented at Position 2.

This is because, regardless how data return order changes, checksum always produces consistent results. As long as we know all the expected data returning from the DDR memory, we can come up with the checksum reference value, to compare against the checksum value produced by the real silicon.

However, CRC values will change per data return order, making them unpredictable with out-of-order data. The exact data return order is purely random in the real system. Even if we know all the expected data returning from the DDR memory, we cannot figure out the CRC reference value without knowing the exact data return order. Therefore, it is only feasible to implement CRC at Position 2, where CRC value is predictable, and the CRC value produced by the real silicon has some reference value to compare against.

Conclusion

We discussed several widely used DFD techniques. Although simple conceptually, silicon debug still requires the validation engineers to understand the system functionality, come up with a comprehensive test plan, closely work with design engineers, and analyze the results carefully.

Q2: How to implement checksum?

Checksum works like an accumulator. Assuming the checksum accumulates incoming 16-bit data on per byte basis, then we can implement a 32-bit checksum like this:

```
logic [31:0] checksum;

always_ff @(posedge clk)
   if (~reset_n)
      checksum <= '0;
   else if (data_valid)
      checksum <= checksum + data[7:0] + data[15:8];
```

Note, checksum could overflow, and it is legal.

Q3: How to implement Cyclic Redundancy Check (CRC)?

Implementing CRC on Serial Bits

We start with CRC implementation on serial bits. First, we need to choose a CRC algorithm or polynomial. In this example, we choose CRC-32, one of the most commonly used CRC algorithms in the industry. For CRC-32, its polynomial is represented by:

$$x^{32} + x^{26} + x^{23} + x^{22} + x^{16} + x^{12} + x^{11} + x^{10} + x^8 + x^7 + x^5 + x^4 + x^2 + x + 1$$

For CRC-32, we need 32-bit flops to implement it. We also need to initialize the CRC register to a known value before any calculation can start. For CRC-32, the initial value is 32'hFFFF_FFFF.

For serial bits, a new data bit comes in every cycle:

1. For CRC register bits whose x index is 0 in the polynomial, they get the value from its previous bit in the chain
2. For CRC register bits whose x index is 1 in the polynomial, they get the XOR of the input data bits and the precious bits in the chain

We show the RTL implementation of CRC-32 on serial bits below:

```
logic [31:0] crc;

always_ff @(posedge clk)
   if (~reset_n)
      crc <= '1;
   else if (init)    // CRC-32 initialization
      crc <= '1;
   else if (data_valid)
      crc <= {crc[30:0], 1'b0} ^
             (32'h04C1_1DB7 & {32{(data_in ^ crc[31])}});
```

Implementing CRC on Parallel Bits

To implement CRC on parallel bits, we essentially repeat the same method for serial bits multiple times, one input bit a time.

Assuming the incoming data has "DATA_WIDTH" bits, we show the RTL implementation of CRC-32 on parallel bits below:

```
logic [31:0] crc, crc_d;

always_comb begin
   crc_d = crc;

   // repeat the same process multiple times per cycle
   for (int i = 0; i < DATA_WIDTH; i++)
      crc_d[31:0] = {crc_d[30:0], 1'b0} ^
                       (32'h04C1_1DB7 & {32{(data_in[i] ^ crc_d[31])}});
end

always_ff @(posedge clk)
   if (~reset_n)
      crc <= '1;
   else if (init)    // CRC-32 initialization
      crc <= '1;
   else if (data_valid)
      crc <= crc_d;
```

Note, CRC on parallel bits may result in a huge XOR tree, if "DATA_WIDTH" is a large value. If we find it hard to close timing, we should consider pipelining the CRC implementation, by breaking and allocating the loop of "DATA_WIDTH" into multiple cycles.

Q4: How to identify which bit of the 32b register has stuck-at fault?

Assume a 32-bit write-only register resides in a black box module, and only 1 bit of the register can have stuck-at fault. You have write access to the register through the 32-bit configuration bus, but you do not have read access.

In addition, you know the bit-XOR value of the 32 bits. How to identify which bit has the stuck-at fault?

Linear Search

The simplest approach is linear search. In the 1st step, you can write 32'h0000_0001 to the register and get the bit-XOR value, and then write 32'h0000_0000 and get the bit-XOR value again. All possible bit-XOR read value combinations are shown in the table below:

Write value	Different scenarios to consider				
	Bit-XOR value w/o stuck-at fault	Bit-XOR value when bit[0] is stuck @0	Bit-XOR value when bit[0] is Stuck @1	Bit-XOR value when bit[31:1] has stuck @0 fault	Bit-XOR value when bit[31:1] has stuck @1 fault
32'h0000_0001	1	0	1	1	0
32'h0000_0000	0	0	1	0	1

Obviously, if we get 0/0 bit-XOR value from the black box, then bit[0] has stuck @0 fault; if we get 1/1 bit-XOR value from the black box, then bit[0] has stuck @1 fault.

If other bits have stuck-at fault, then the bit-XOR value is either 1/0 or 0/1. We should repeat this process to check all remaining bits.

In the worst case scenario, we will need 32 x 2 = 64 patterns to check the stuck-at fault using linear search.

Binary Search

Although intuitive, linear search takes too many iterations to get the final results in the worst case scenario. An optimized solution is to use binary search.

In the 1st step, you can write 32'hFFFF_FFFF to the register and get the bit-XOR value, and then write 32'hFFFF_0000 and get the bit-XOR value again. All possible read value combinations are shown in the table below:

Write value	Different scenarios to consider				
	Bit-XOR value w/o stuck-at fault	Bit-XOR value when bit[15:0] has stuck @0	Bit-XOR value when bit[15:0] has Stuck @1	Bit-XOR value when bit[31:16] has stuck @0 fault	Bit-XOR value when bit[31:16] has stuck @1 fault
32'hFFFF_FFFF	0	1	0	1	0
32'hFFFF_0000	0	0	1	1	0

If we get 1/0 bit-XOR value from the black box, then bit[15:0] has stuck @0 fault; if we get 0/1 bit-XOR value from the black box, bit[15:0] has stuck @1 fault. Either case, we shall continue the binary search for bit [15:0].

If bit[31:16] has stuck-at fault, the bit-XOR value is either 1/1 or 0/0, then we shall continue the binary search for bit[31:16].

In the 2nd step, the search range is cut in half. Obviously, we will need log2(32) = 5 steps, and 5 x 2 = 10 binary search patterns to check the stuck-at fault in the worst case scenario.

As you can see, the binary search method greatly improves the iteration time and reduces the number of patterns needed.

Part 3 Behavioral Questions & Useful Interview Tips

Q1: What verbs to use in your resume to stand out?

Resumes make the first impression on HR and hiring managers. Remember to use powerful verbs to make your resume stand out.

If You Led a Project

1. Controlled
2. Coordinated
3. Executed
4. Headed
5. Operated
6. Orchestrated
7. Organized
8. Oversaw

If you Created or Developed a Project

1. Administered
2. Built
3. Charted
4. Created
5. Defined
6. Designed
7. Developed
8. Founded
9. Engineered
10. Established
11. Formalized
12. Formed
13. Formulated
14. Implemented
15. Incorporated
16. Initiated

17. Instituted
18. Introduced
19. Launched
20. Pioneered

If You Saved Company Resources

1. Conserved
2. Consolidated
3. Decreased
4. Deducted
5. Diagnosed
6. Lessened
7. Reconciled
8. Reduced
9. Yielded

If You Increased Efficiency, Sales or Revenue

1. Accelerated
2. Achieved
3. Advanced
4. Amplified
5. Boosted
6. Capitalized
7. Delivered
8. Enhanced
9. Expanded
10. Expedited
11. Generated
12. Improved
13. Maximized
14. Outpaced
15. Stimulated
16. Streamlined

17. Strengthened
18. Updated
19. Upgraded

If You Managed a Team

1. Aligned
2. Directed
3. Enabled
4. Facilitated
5. Fostered
6. Formed
7. Guided
8. Hired
9. Inspired
10. Mentored
11. Motivated
12. Recruited
13. Regulated
14. Shaped
15. Supervised
16. Trained
17. Unified
18. United

If You Research, Study or Analyze

1. Analyzed
2. Assembled
3. Assessed
4. Audited
5. Calculated
6. Discovered
7. Evaluated
8. Examined

9. Explored
10. Forecasted
11. Identified
12. Interpreted
13. Investigated
14. Measured
15. Qualified
16. Quantified
17. Simulated
18. Surveyed
19. Tested
20. Tracked

You Achieved Something

1. Attained
2. Awarded
3. Completed
4. Demonstrated
5. Earned
6. Exceeded
7. Outperformed
8. Reached
9. Showcased
10. Succeeded
11. Surpassed
12. Targeted

Conclusion

You might want to review your resume again, and switch the plain boring verbs with these powerful ones.

Q2: Interview etiquette and best interview practices

Before the Interview

1. Become familiar with the company, the team, the position and the interviewers, if possible. This will ensure there is never a lull in the conversation
2. Organize your resume. Make sure you can tell a reasonable and complete story for every single bullet point in your resume
3. Turn off your phone before the interview
4. Never be late for an interview. Give yourself an extra 15 minutes before the scheduled interview time

During an Interview

1. Wait for your interviewer to initiate the handshake. Make sure your grip is firm but not crushing
2. It will be ideal to have an interactive conversation between you and your interviewer, but this may not always be the case. Regardless, you should not be having a one-sided conversation
3. Do not cross your legs when sitting. Body language is crucial in an interview, so maintain a good posture at all times
4. A smile goes a long way, and smile even when you are nervous.
5. Remember your manner, and use words like "please" and "thank you" frequently
6. Never chew gum, candy, mints or chocolate during your interview
7. Use only the name the interviewers use to introduce themselves
8. When the interviewer asks you if you need to use the restroom, always say yes. This gives yourself some time to refresh
9. Politely ask the hiring manager how the hiring process looks, and how long you should expect a final decision from them. This is useful for following up with your job application later on

If You Are Invited to an Interview Lunch

1. This typically happens during an on-site interview. Do not expect that you can enjoy a full lunch, and the interviewer may keep asking you questions during the lunch
2. Even if your interviewer offers you a drink, politely refuse and stick to water during the lunch
3. Take small bites of food, and never talk with your mouth full
4. Stick with food that is eaten with a fork and knife, and not with fingers (like wings, chicken legs, etc.).
5. No matter how delicious it may sound, stay away from the most expensive items on the menu. A safe option is, order the same thing as what your interviewer order for lunch
6. If you are invited to the meal, you are not obligated to pay. Offering payment would be inappropriate

After the Interview

1. Always send a thank you note as soon as you can after the interview. The format of the note may depend on the type of the company you interviewed with, your strengths you want to highlight, or some further clarifications for the answers you gave during the interview
2. No matter how badly you may want to update your social media status, do not share details of your interview on the Internet. Your interviewer may be checking up on you online

Q3: How to answer "Any Questions for Me" at the end of an interview?

An interview runs in two ways: while the interviewer is testing you and evaluating your potential, you should also assess whether the company or the team is suitable for you. Therefore, you should ask questions that are relevant to your interests and are important to you.

The following are a few sample questions that you can ask towards the end of an interview.

When do I expect to get updates of my job application?

This is the chance for you to understand the hiring process and timeline. In addition, it can be useful for following up with your job application later on, and gives you an idea when it is reasonable to follow up if you never hear back from the HR or the hiring manager.

What is your daily workflow?

This is a relatively basic question that many interviewees will ask.

What motivates you to work here every day?

Common answers include opportunities for learning and growth, collaboration, or the product. It gives you an idea of job satisfaction and an opportunity to see if this position is a cultural fit for you.

What kind of team building activities do you have, and how often?

This question tends to give you an idea of team culture, and a hint of work life balance.

How many people are currently in your team? What is the distribution between senior and junior engineers?

This question helps to gauge whether there might be too few or too many people in the team, depending on your job-seeking needs.

What is the relationship between RTL designers and architects / DV engineers / FV engineers / physical design engineers?

This question gives you a general idea of cross functional team interactions, like who drives the decision making process, and who holds responsibility when a specific issue occurs.

If I join the team, what will be my immediate responsibility? Which area can I immediately contribute to?

This question helps you understand whether your strengths align with the interviewer's requirements.

It also allows you to further promote yourself. If the interviewer expresses interest in your expertise in certain areas, you can emphasize your interest in those areas to leave a positive impression. When writing a thank-you note, you will also know which aspects of yourself to highlight.

What expectations do you have for me in 1 or 2 years?

This question gives you an idea whether the management has considered their direct reports' long-term career path. If they do, this could be an ideal company / team you want to work for.

During a performance review, what do you value the most?

This question gives you a clearer understanding of the evaluation system and promotion path. Again, remember to highlight your strengths.

Q4: What to write in a "Thank You Letter"?

Addressing Concerns

If the interviewer mentioned specific criteria during the interview that you felt you fall short of, address this again in the thank you letter. Briefly reiterate the point, emphasize why it would not be a concern for you to excel in the position.

Rectifying Missed Points

Reflecting on the interview, you might realize certain points you did not mention. Use this chance to elaborate on those topics, expressing your desire to provide further insights.

Expressing Gratitude

There are instances where you simply want to express gratitude for the opportunity. However, do not miss the chance to reinforce your candidacy by highlighting your qualifications, expressing your strong interest in the company, and reaffirming your commitment to achieving their goals.

Keeping it Concise

Although the thank you note serves as a sales tool, it does not need to be as long as your resume or cover letter. Keep it concise while effectively leveraging the information you got during the interview to strengthen your candidacy.

Q5: Why and how should you follow up with your job application?

Following up with your job application is critical in showing your continued interest in this job opportunity. In addition, it can give you peace of mind.

For example, the HR or the hiring manager told you that you would hear back by a certain date, but that date has passed and you have not heard anything. You may start to panic and over analyze the situation. Following up may get you an update from the HR or the hiring manager, and you could be relaxed a little.

It is recommended to wait at least 3 business days after the aforementioned date by which you expect to get updates, and tastefully follow up with your job application.

The follow up should be concise and friendly, and you may phrase your message along the lines of, "I know you mentioned you planned to make the final decision by the end of this month, and I wanted to follow up and see where you are in that process".

Follow up can either move things along or give you closure. If it results in an offer, that is great. If you do not receive any response from the company, follow up once more and then move on. Either way is good.

However, if you are following up multiple times after an interview, that is quite unlikely appreciated.

www.ingramcontent.com/pod-product-compliance
Lightning Source LLC
Chambersburg PA
CBHW071102240526
45471CB00016B/2398